ULTIMATE SUPERCARS

AUDI R8

By Meg Greve

Kaleidoscope
Minneapolis, MN

The Quest for Discovery Never Ends

....

This edition first published in 2023 by Kaleidoscope Publishing, Inc.

No part of this publication may be reproduced in whole or in part without written permission of the publisher.

For information regarding permission, write to
Kaleidoscope Publishing, Inc.
6012 Blue Circle Drive
Minnetonka, MN 55343

Library of Congress Control Number
2022937980

ISBN
978-1-64519-606-8 (library bound)
978-1-64519-676-1 (ebook)

Text copyright © 2023 by Kaleidoscope Publishing, Inc. All-Star Sports, Bigfoot Books, and associated logos are trademarks and/or registered trademarks of Kaleidoscope Publishing, Inc.

Printed in the United States of America.

FIND ME IF YOU CAN!

Bigfoot lurks within one of the images in this book. It's up to you to find him!

TABLE OF CONTENTS

Chapter 1: Daydreaming About a Legend **4**

Chapter 2: Thanks, Mr. Horch! **12**

Chapter 3: The R8 in Numbers **18**

Chapter 4: Speeding into Action! **24**

Beyond the Book *28*
Research Ninja *29*
Further Resources *30*
Glossary *31*
Index *32*
Photo Credits *32*
About the Author *32*

Chapter 1
Daydreaming About a Legend

Aiden draws a car on top of his math paper. It is sleek, and low to the ground. He imagines he is a sportscaster and mumbles, "Like a lion on a hunt, the engine roars to life. The sound rumbles in your belly. The driver hits the gas. The car shoots forward!"

"Aiden," says a stern voice. "What is the answer to the first problem?"

Aiden's head snaps up. "Uh, R8?"

"Aiden, you can stay after school to finish your work," sighs his teacher.

FUN FACT
Hypercars can cost $850,000 or more!

Since he first learned about supercars, the Audi R8 has been Aiden's dream car. Most supercar fans would not blame him. The R8 is like driving a high-speed race car with all the comforts of a luxury car.

Supercars like the Audi R8 are designed in a sports car style and are legal to drive on the street. They have features making them cost $150,000 or more. The top-performing supercars are called hypercars.

PARTS OF AN
AUDI R8

front bumper

tires

steering wheel

vertical panels

The Audi R8 uses space frame technology. This means that most of the frame is made of aluminum. This metal makes the car lighter.

Audi R8 C

engine

rear wing

back bumper

exhaust pipes

This two-seater sports car can have an all-wheel drive or rear-wheel drive system. It has a mid-engine that is behind the seats. This keeps the R8 balanced when hugging roads at higher speeds. The front end allows extra crush space if the car has a head-on **collision**.

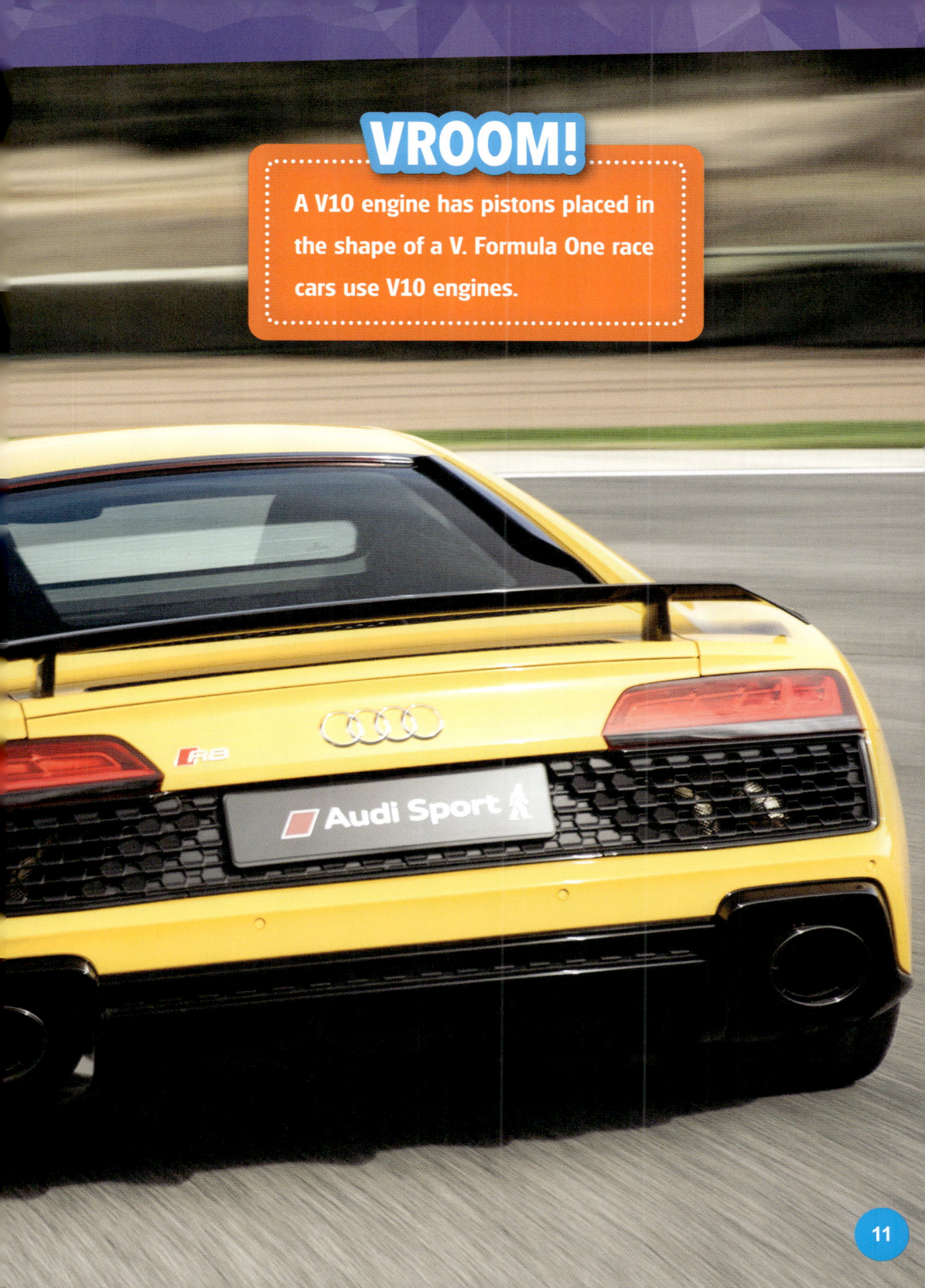

VROOM!

A V10 engine has pistons placed in the shape of a V. Formula One race cars use V10 engines.

Chapter 2
Thanks, Mr. Horch!

Audi was founded in Germany by August Horch and three other car companies known as Horch, DKW, and Wanderer. They formed a group called Auto Union in 1932.

The company we know as Audi today began in 1969. They created the R8 design using the Le Mans **concept** car that won many races in the early 2000s. In 2006, production started on the car that would become the dream of car owners around the world. As of 2022, there are eight different models of the R8.

AUDI R8 TIMELINE

2007

2007–2012:
Audi R8 V8

2008

2008–2012:
Audi R8 V10

2012

2012–2015:
Audi R8 V10

Audi R8 V8

2015

2015–2018:
Audi R8

2017

2017–2021:
Audi R8 V10 RWS

2018

2018–Present:
Audi R8 Coupe

2021

2021–Present:
Audi R8 V10
Performance RWD

The **headquarters** of Audi is in Ingolstadt, Bavaria, Germany. Audi sells over 1.6 million cars a year. In 2021, only 648 lucky people bought R8s.

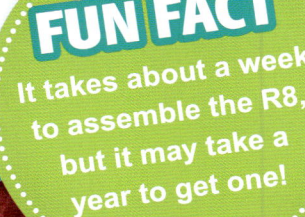

FUN FACT
It takes about a week to assemble the R8, but it may take a year to get one!

WHERE THE AUDI R8 IS MADE

MADE IN NECKARSULM

Audis are made in 9 different places around the world. They only assemble the R8 in Neckarsulm. It takes up to 70 workers to assemble 5,000 parts by hand.

If you are able to buy an Audi R8, you have many choices to build your dream car. There are two different styles—the coupe and the performance Spyder, which is the **convertible**. You can choose the outside paint color, the style, the wheels, the exhaust system, and even the stitching color on the leather seats!

FUN FACT
The word quattro means four in Italian, which is a clever name for all-wheel drive!

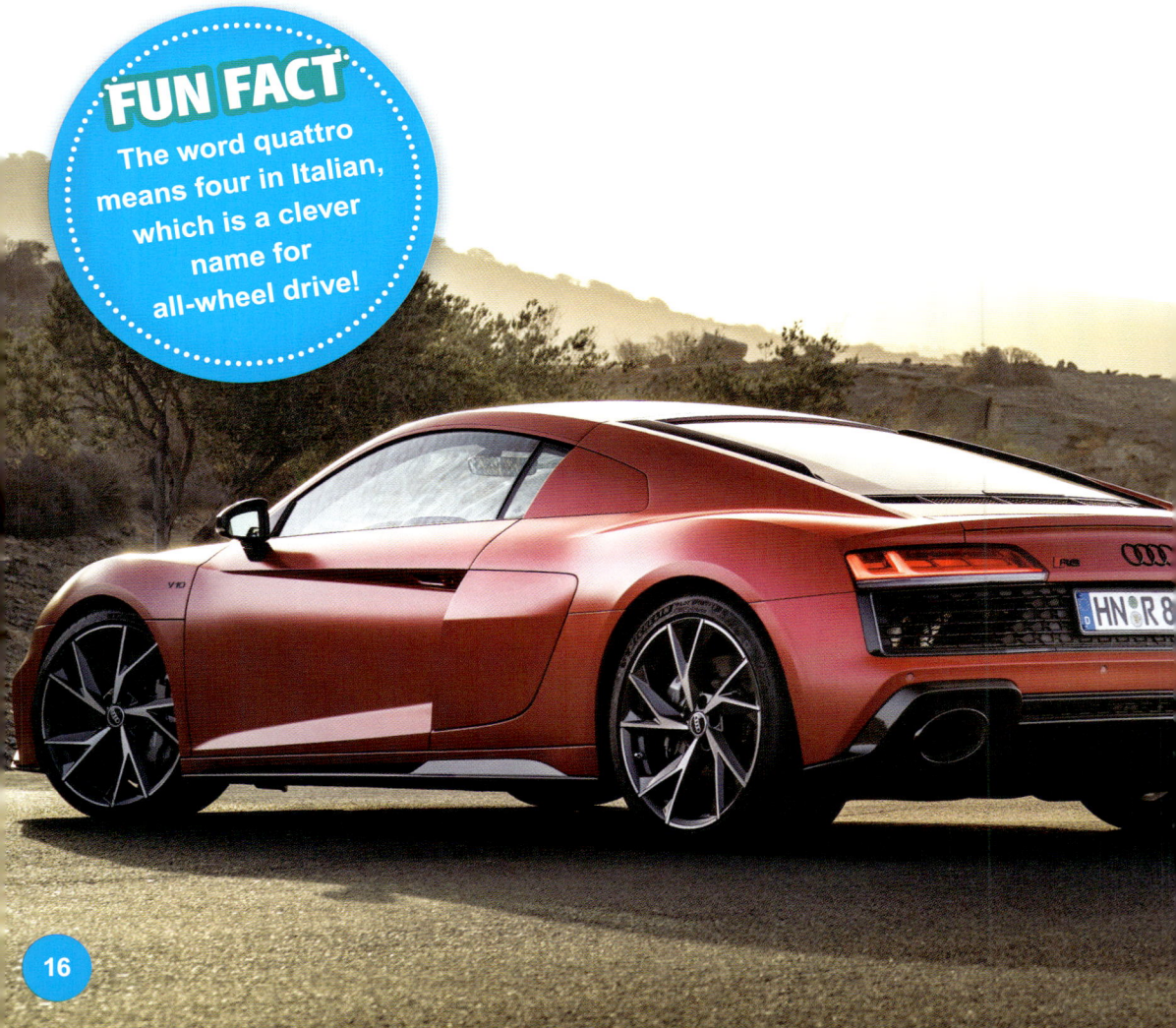

RWD or performance quattro? RWD, rear-wheel drive, means all the power from the engine goes to the back wheels to push the front ones. RWD makes the car less expensive and lighter than a quattro. The quattro is all-wheel drive. It is heavier but will grip the road better.

LISTEN TO THIS

Horch means listen in German. In Latin, listen is *audi*. The Audi logo has four rings. Each ring stands for one of four car companies of the Auto Union.

Chapter 3
The R8 in Numbers

The R8 is meant for people who love driving. However, it does not have many fancy features. For example, there is a Bentley that has a starlit sky as its ceiling!

The R8 is for the driver who wants to have fun. There is a state-of-the-art sound system, leather seats, and a dash view that shows information about the car. The best part is feeling like you are driving in a high-speed race.

FUN FACT
Tony Stark the character in the *Marvel Movies*, drove an R8 because the script called for him to drive something sporty.

THE AUDI R8 IN DETAIL

COST: starts at about $150,000

Height: 4 feet high (1.2 m)

Width: 6.7 feet wide (2.0 m)

LENGTH: 14.5 feet long (4,426 mm)

WEIGHT: up to 3,792 pounds (1,720 kg)

TOP SPEED: Coupe – 204 miles per hour (328 km/h)
Spyder – 203 miles per hour (327 km/h)

TIME FROM 0-60 miles per hour (96 km/h): 3.7s (Coupe)
3.8s (Spyder)

The **aerodynamics** of a car can change its speed. Air flows over and around a car faster than under it. The R8's rear **spoiler** will stick out when the back of the car lifts. It pushes the car down, blocking the air moving over it.

Brake calipers squeeze the brake pads against the rotor to stop the car. The color of the caliper does not affect its performance, but it does add a nice detail.

Once you buy your R8, you can add things to make it wider, longer, or taller. Changing exhaust pipes can improve the car's ability to pick up speed. You can change out the brakes and add in colors to stand out.

Chapter 4
Speeding into Action!

The Audi R8 is a favorite in the car world. It has the same look as race cars that have winning records. The style of the car is not flashy, but it's exciting to drive.

After the 2022 model is presented, Audi will try something new for the R8. The goal is to create an electric model. **Emissions** from cars have hurt our climate. Electric cars help limit pollution and make the planet healthier.

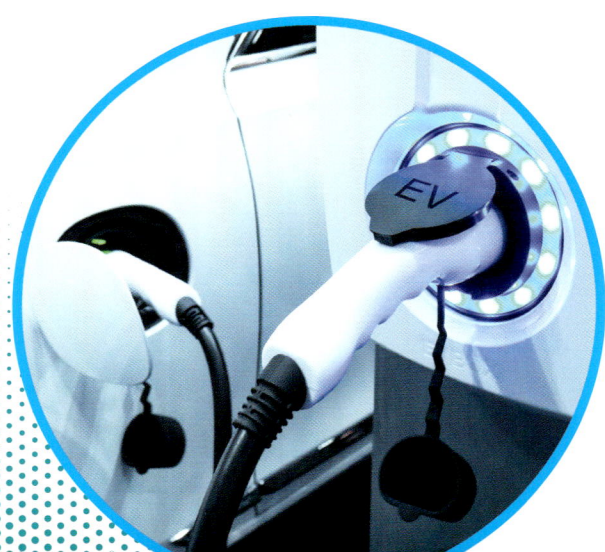

The bell rings. Aiden makes his way back to his math teacher's room. He can't wait until the auto show this weekend. There will be hundreds of cool cars. But he has his heart set on seeing the Audi R8. Its body is shaped like a rocket. When it is turned on, it sounds like one too.

When he is old enough to own an R8, Aiden knows it will be fully electric. The V10 engine will be a thing of the past. But it will still be his favorite supercar, and he can't wait to drive one!

FUN FACT
Scientists and others believe that electric cars are part of the solution to reduce pollution that will help slow climate change and global warming.

BEYOND
THE BOOK

After reading the book, it's time to think about what you learned. Try the following exercises to jump-start your ideas.

THINK

FIND OUT MORE. The R8 is built only in Germany. There are nine manufacturing plants around the world. Research some of the other plants. Where are they located? Which cars are built there? Create a map showing where these places are located.

CREATE

GET ARTISTIC. Use a primary resource, such as a photo, interview with an owner, or original document, to learn more about the Audi R8. What new information do you want to learn about the R8? How can you locate the information?

SHARE

DIG DEEPER. Create a sales brochure for the Audi R8. What important information would you include in the brochure? What images would you use that would interest your buyer?

GROW

GO TO A CAR SHOW. With an adult, see if you can visit a car dealership that carries the R8 or other supercars. What would you hope to see or learn about the car?

RESEARCH NINJA

Visit www.ninjaresearcher.com/6068 to learn how to take your research skills and book report writing to the next level!

RESEARCH

DIGITAL LITERACY TOOLS

SEARCH LIKE A PRO
Learn about how to use search engines to find useful websites.

FACT OR FAKE?
Discover how you can tell a trusted website from an untrustworthy resource.

TEXT DETECTIVE
Explore how to zero in on the information you need most.

SHOW YOUR WORK
Research responsibly—learn how to cite sources.

WRITE

GET TO THE POINT
Learn how to express your main ideas.

PLAN OF ATTACK
Learn prewriting exercises and create an outline.

DOWNLOADABLE REPORT FORMS

Further Resources

BOOKS

100 Cars That Changed the World: The Designs, Engines, and Technologies That Drive Our Imagination. Morton Grove, IL: Publications International, Ltd., 2020.

Built for Speed: World's Fastest Road Cars. Morton Grove, IL: Publications International, Ltd., 2019.

Lamm, John. *Supercar Revolution: The Fastest Cars of All Time.* Minneapolis, MN: Motorbooks, 2018.

WEBSITES

Factsurfer.com gives you a safe, fun way to find more information.

1. Go to www.factsurfer.com.
2. Enter "Audi R8" into the search box and click 🔍
3. Select your book cover to see a list of related websites.

Glossary

aerodynamics: the way air moves around objects. An aerodynamic design makes a car move faster. The rear spoiler of the Audi R8 holds the car down to make it more aerodynamic.

collision: a crash between objects. In head-on collisions, people riding in an R8 have more protection than in many other cars. The empty space allows for extra crush space.

concept: something that is a new idea. The Le Mans concept car won many races before it was in production.

convertible: a car that has a roof that can be removed. The Audi R8 performance Spyder is a convertible.

emissions: chemicals in the exhaust that is pollution. After the 2022 R8, Audi's plan is to make a fully electric R8 that will have fewer emissions and be better for the planet.

headquarters: the center or main part of a company. Audi's headquarters is in Ingolstadt, Bavaria, Germany.

luxury: a state of extreme wealth and comfort. The Audi R8 gives you the driving experience of a race car with the comforts of a luxury car.

spoiler: a piece of the car that is attached to the back end, changing airflow around the car. The R8's spoiler is unique because it pops up out of the back when the car gets too much lift to push it back down.

Index

aerodynamics, 22
convertible, 16
cost, 7
coupe, 16
drive, 7, 10, 17, 18, 24, 27
engine, 4, 10, 17, 27
Horch, August, 12

hypercars, 7
model(s), 12, 24
quattro, 17
spoiler, 22
supercar(s), 6, 7, 27
wheel(s), 10, 16, 17

PHOTO CREDITS

The images in this book are reproduced through: Roman.S/Shutterstock 3; BoJack/Shutterstock 13; Yauhen_D/Shutterstock 14; twilllll/Shutterstock 19 (background); youness maach/Shutterstock 21; FranciscoMarques/Shutterstock 22; VanderWolf Images/Shutterstock 22-23; buffaloboy/Shutterstock 24; sippakorn/Shutterstock 25; Teddy Leung/Shutterstock 30. All other images courtesy Audi Media Center (Graeme Fordham 4-5; sagmeister_potography 12, 16-17). **Cover:** Courtesy of Audi Media Center, Maciej Bledowski/Shutterstock (background).

About the Author

Meg Greve lives in Chicago, where it is not uncommon to see supercars on the road. She has two children and a husband, so she needs something bigger than the Audi R8! She dreams one day of taking a ride in that fancy car on the open road somewhere outside of the city.